BUG HUNTERS

ALSO BY ADA AND FRANK GRAHAM

Whale Watch

AN AUDUBON READER

BUG HUNTERS

ADA AND FRANK GRAHAM
ILLUSTRATIONS BY D. D. TYLER

DELACORTE PRESS/NEW YORK

Published by
Delacorte Press
1 Dag Hammarskjold Plaza
New York, N.Y. 10017

Designed by Leo McRee

Manufactured in the United States of America

First printing

Library of Congress Cataloging in Publication Data

Graham, Ada.
    Bug hunters.

    (An Audubon reader)
    Bibliography: p.
    Includes index.
    SUMMARY: Describes the biological control method
of curbing harmful insects in which man uses
nature itself to fight nature's pests.
    1. Insect control—Biological control—Juvenile
literature. [1. Insect control] I. Graham,
Frank, 1925–   joint author. II. Tyler, D. D.
III. Title. IV. Series.
SB933.3.G73   632'.6   77-20531
ISBN 0-440-00909-X
ISBN 0-440-00910-3 lib. bdg.

To the memory of Marie Rodell,
conservationist and helpful friend
to so many of us who write about the natural world,
this book is dedicated.

CONTENTS

1
A CURIOUS HARVEST

It was a warm, hazy June morning in the rolling farm country of northern Indiana. The climbing sun had spread a glimmering whiteness across the east, while thunderclouds began to mass darkly half a sky away. A dozen or more men and women walked into a field that was hedged by poplars and thick bushes and planted with oats. Each member of the party carried clippers and a large round container.

They set to work among the growing oats at once, gathering a curious harvest. They clipped handfuls of the emerald green leaves and set them upright in the containers. Once in a while they stopped to slap mosquitoes which had added their irritating whine to the already heavy air.

"We'll have to work fast," said Dr. Thomas Burger,

one of the members of the group, as he glumly studied the sky. "Those little planes will be lucky to get off the ground once it starts to storm."

Burger slapped at another mosquito, but it was not the only insect on his mind that morning. He is a stocky young man, energetic and decisive in his movements. He is the director of a laboratory operated by the United States Department of Agriculture in Niles, Michigan, not far from the Indiana grainfield where he worked this morning.

The Department of Agriculture—often called USDA —is the government agency that (among other things) looks after the interests of American farmers. One of the USDA's most important jobs is to help farmers fight the insect pests that threaten to destroy their crops. For many years the USDA relied mostly on deadly chemical sprays called pesticides to do the job. But pesticides, while they kill many insects, have not solved the problem. Not only are they very expensive, but they also poison many helpful animals—and even human beings—yet the pest insects keep returning every year anyway.

Tom Burger is one of the USDA scientists who is working for new and safer ways to get rid of pests that attack our farms and forests. Like Burger, the other men and women in this grainfield worked for federal or state agencies concerned with agriculture. They had come to collect the oat leaves, which were already covered with the eggs and the larvae (or young) of the cereal leaf beetle.

The adults of these beetles lay their eggs on the leaves of oats, barley, wheat, and corn. When the eggs hatch, the larvae and later the adults begin to feed on the leaves. These insects chew their way through entire grainfields. Sometimes they cause so much damage that the crop is ruined.

Burger and his helpers were packing their containers with the oat leaves and their cargo of pest insects to deliver them to a nearby airport. From there, small planes would distribute the leaves and beetles to farm communities throughout Indiana and its neighboring states. It is part of a plan to save the crops of Midwestern farmers.

"The leaves through here are full of the larvae," Burger said, pointing to the rows of growing oats. "See, they've got that frosted look to them."

Burger's companion bent to look at the oat leaves. He could see some of the eggs of the cereal leaf beetle. They were clear yellow cylinders laid one by one on the upper surfaces of the leaves, and they stood out against the bright green oats though they were less than one-sixteenth of an inch long. Some of the eggs were growing darker with age.

But there were many more larvae than eggs. The larvae, already three-sixteenths of an inch long, also looked dark until Burger rubbed away the brown excrement with which they cover themselves to keep from drying out. Then his companion could see that the fleshy, slug-like bodies of the larvae were yellowish with a black head and legs. A few of the dark-red and blue-black adult beetles, about the size of the larvae, were also on the leaves.

The oat leaves, as Burger had pointed out, looked "frosted." The larvae do not eat completely through a leaf as the adults do. They simply eat long strips of green between the leaf's veins so that its white lower epidermis or skin remains, making the leaf look as if it were touched with frosting.

The workers had finally filled their containers with leaves. They covered them with moist cloth, sealed them with tape, and stacked them in waiting trucks. Tom Burger rode in one of the trucks to the airport. He and his helpers worked fast to load the small planes that were owned by the state of Indiana. There was still time to fly the leaves to farm communities in Indiana before the storm struck. Trucks rushed the fresh leaves to nearby states where heavy rain made flying dangerous.

"Nothing is ever easy about this job," Tom Burger said. "We seem to have problems every step of the way, but we are going to make this program a success."

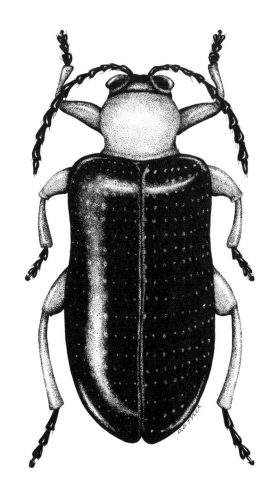

A cereal leaf beetle.

Other workers were ready in the farming communities to receive the containers of oat leaves. Later in the day they distributed the containers to local farmers. The farmers, in turn, carried the oat leaves into their fields and stood them next to their own grain plants. When the oat leaves wilted, the larvae would move onto the farmers' plants. By that evening, fresh supplies of the eggs and larvae of the destructive cereal leaf beetle were spread throughout nearly every grain-growing area in the Midwest.

Fiendish, yes. But the beetles and not the farmers were going to be the victims of this plot. It had been planned for a long time. Burger and the other federal and state agricultural scientists had chosen this eight-acre field because it was an ideal place for cereal leaf beetles. The field was planted with oats—a favorite food of those insects. It was surrounded by dense bushes which gave the insects plenty of cover during the winter.

The oat field, then, is an insectary—a place where insects are reared. Yet rearing beetles was only a beginning. The most important part of the plan was to produce a nasty surprise for the beetles themselves.

The eggs and larvae that Tom Burger shipped out of the insectary on oat leaves that morning carried a deadly cargo. Inside many of them was a tiny wasp that even then was growing rapidly by eating the tissues of its "host"—the animal on which it feeds. The eggs and young of the cereal leaf beetle were in reality carrying the seeds of their own destruction many miles from the insectary.

Tom Burger is an expert in biological control, an exciting method of curbing pests. In biological control, people learn to use nature itself to fight nature's pests. Burger and his helpers were taking advantage of the fact that certain tiny wasps lay their own eggs in the eggs or larvae of the cereal leaf beetle.

As the tiny wasp grows inside the egg or larva of the beetle, it goes on eating its host and finally kills it. Then it eats its way out and begins its life in the farmer's grainfield. When it comes time to lay its own egg, this wasp searches for the eggs and larvae of other cereal leaf beetles. Only there will it lay its egg. Its young, in turn, will grow by feeding on the young of the destructive beetle. From year to year new generations of these wasps help to carry on the USDA's crusade against the cereal leaf beetle.

This program suggests how much people might really do to create a better environment in the right way. To understand how this can be done, we must understand the part biological control plays in the struggle against destructive plants and insects.

2
PREDATORS AND PARASITES

The British naturalist and writer H. M. Tomlinson once described a praying mantis, a large insect which a friend of his kept as a pet. His friend called it Edwin:

"Edwin has a long thin neck—the stalk to his wings as it were—which is quite a third of his length. At times he will remain still, with his hands clasped up before his face, as though in earnest prayer for a trying period. If a fly alights near him he turns his face that way and regards it attentively. Then sluggishly he approaches it for closer scrutiny. Having satisfied himself it is a good fly, without warning his arms shoot out and that fly is hopelessly caught in the hooked hands. He eats it as you

A praying mantis is a predator that feeds on many kinds of insects.

do apples, and the mouthfuls of fly can be seen passing down his glassy neck."

Like many insects, the praying mantis is entertaining and instructive to watch as it goes about its affairs. But perhaps its greatest practical value to us lies in its ability to catch and eat flies and other pest insects. Someone, many centuries ago, must have looked at such a hungry insect as the praying mantis and decided to put it to work for the good of mankind. That was the beginning of biological control.

Putting insects to work eating other insects has been called "the thinking person's pest control." It is an old science. In ancient times Chinese farmers used various ants to control leaf-eating insects on their fruit trees. The Chinese even built little bridges of bamboo poles to help the ants move from tree to tree.

Most insects are either vegetarians (herbivores) or meat-eaters (carnivores). Just as cows eat grass and tigers eat other animals, so some insects eat only plants, while others eat only the plant-eating insects.

The insect-eating insects are usually divided into two kinds, predators and parasites:

A *predator* captures and eats a large number of insects of all kinds. The praying mantis is a predator. It eats flies and almost any other creature it can catch. There are even reports of a praying mantis catching small birds such as hummingbirds.

Predators and Parasites

A *parasite*, as it grows to maturity, usually eats only a single animal, which is called its host. The adult parasite lays its eggs on, or inside, the body of the host animal or insect. Certain tiny wasps and flies lay their eggs inside the bodies of caterpillars, which is the common name for larvae. When the young hatch, they begin feeding on the unfortunate caterpillar and eventually kill it. By that time, the young parasites have grown to maturity. They leave their host and go out to mate and lay their own eggs in some other caterpillar.

Insects have been on earth for a very long time. Their remains have been found in rocks that are more than three hundred million years old. The word *insect* comes from the Latin word for *incised*, which means "deeply cut" and refers to the sharp divisions between each of the mature insect's three parts—the head, the thorax, and the abdomen.

An insect has no bones to shape its body, as we do. Instead, it wears its skeleton—or hard covering—on the outside of its body, as lobsters and crabs do. It has three pairs of legs and usually two pairs of wings, which are attached to the thorax. Spiders and mites, which have four pairs of legs, are not insects.

Many kinds of insects grow in a series of four stages called metamorphosis. The first stage is the egg. Then the young insect, or larva, hatches from the egg. The larva, which is usually wormlike, must keep shedding its hard skin or skeleton as it grows. When it is fully grown, it has a resting stage during which it becomes a pupa. It may spend its time as a pupa in a cocoon or a pupal case which it makes to cover itself.

Finally, the pupa changes into an adult, acquiring wings or other parts that make it look very different from what it did as a larva. The life cycle of the monarch butterfly is a good example of complete metamorphosis: (1) the egg, (2) the larva or caterpillar, (3) the pupa or chrysalis that does not spin a cocoon, and (4) the adult butterfly.

Entomologists, or scientists who study insects, tell us they have identified nearly one million kinds of insects in the world. There are probably several million other kinds they have still not identified. Yet only a very small percentage of the world's insects are serious pests to people —perhaps less than one percent. The others are harmless or, as in the case of the honey bee or the praying mantis, helpful to us.

Not all pests are insects, either. Many kinds of plants are serious pests, such as the weeds that choke out valuable crops on farmland. Certain kinds of fish become pests, such as the sea lamprey which destroyed many of the whitefish and lake trout in the Great Lakes. Some mammals, for example, certain rats and mice, are pests around homes and warehouses.

Predators and Parasites

Most insects are not pests. They live in harmony with their environment, and are usually kept from becoming pests by the many predators, parasites, and diseases that naturally attack them. But because of their great number and their ability to reproduce quickly, insects can become more prominent pests. The expert in biological control studies the life cycle of a pest to discover its most vulnerable stage and then tries to find a predator or a parasite to attack those weak points.

Insects that are otherwise harmless often become a problem when they somehow are transported to a new country. There they may not find the natural enemies that kept them under control in their native country. Then, they have a "population explosion" and become pests. Many of the famous cases of successful biological control took place when farmers or entomologists were faced with a serious pest that had entered their country from abroad. To solve the problem they traveled to the country where the insect came from and brought back some of its natural enemies. It seems that the first successful case of this kind took place in 1762. In that year, the people of India imported myna birds to feed on the red locust, which had become a pest on their farms.

Entomologists often define biological control "as the importation and use of natural enemies to control insect pests and weeds." For instance, when a wood wasp began to damage pine trees in southern Australia, entomologists searched eighteen countries in Europe and North Africa to find predators and parasites that would control the wood wasps.

Many years ago a poisonous plant called the Klamath weed somehow was imported from Europe to the western United States, where cattle ate it and became sick. American scientists imported two beetles that feed on this weed in Europe. Within fifteen years, the beetles destroyed so much of the Klamath weed that it was no longer a pest. Perhaps scientists can find a bug that eats poison ivy!

One of the strangest cases of biological control involves some beetles that eat dung or manure. When Europeans settled Australia—which had very few large mammals—they imported cattle, horses, sheep, and other large domestic animals. As the settlers' herds of livestock increased, the animals littered the fields with dung. On other continents, dung-eating beetles clean up the fields, but in Australia the dung just lay there because no beetles had evolved to take care of that job. The dung smothered the grasses and provided places for pesky flies to breed.

Australian entomologists solved the problem by going to Europe and Africa where there are many kinds of beetles that feed on the dung of large mammals. They imported them to Australia and the beetles soon cleaned up the fields.

A giant dung beetle on manure.

3

BUG HUNTERS

Paul DeBach is an entomologist, writer, and editor who has played a large part in developing biological control in the United States. He recalls his days as a sophomore at the University of California at Los Angeles when he attended a series of lectures given by an expert on biological control, Professor Harry S. Smith.

"This was the turning point in my professional career," Paul DeBach writes. "Biological control was so intellectually satisfying, so biologically intriguing and so ecologically rational that I immediately decided to become a specialist in this field. This I did and I have always been happy with my choice."

DeBach might have added that the career he chose is not only satisfying to the mind but it has moments of high excitement and adventure. He has traveled to distant parts of the earth to search for predators and parasites that may control pest insects. He has collected insects in jungles and deserts. He has had to put up with dangers and discomforts. In his excellent book *Biological Control by Natural Enemies*, he writes:

"I have had to have armed guards in order to travel and collect in certain localities in Burma; have lived in hovels on stilts where the toilet was a hole in the floor; have had pig-intestine soup for what I later recalled was Thanksgiving Day dinner; have had to have local permission to enter and collect in bandit-infested parts of the Tribal Territories of West Pakistan. On the other hand, I have had some lovely collecting in Rio, Tahiti, Japan and other places."

Paul DeBach is a modern scientist who is carrying on in the footsteps of earlier Americans who were pioneers in biological control. The pesticide industry today boasts that agriculture would not be possible without the use of poisons to kill pests. But American agriculture became the wonder of the world long before the modern pesticides were invented. This was largely because of the help it received from the natural enemies of pests.

The settlers, as they traveled westward across the American continent, had to battle some native pests. The story of the Mormons and the gulls is a famous incident in American history. In 1848, when the Mormons' new colony in Utah was struggling to harvest its first food crop, the fields suddenly were overrun with wingless insects that are now called Mormon crickets.

Hordes of crickets and their relatives, the locusts, have caused great disasters in many parts of the world. They move through planted fields, devouring the plants and leaving the earth barren behind them. The Mormons were trying to build a new life in a harsh and isolated land. A historian of Utah tells what happened:

"Just in the midst of the work of destruction great flocks of gulls appeared, filling the air with their white wings and plaintive cries, and settled down on the half-ruined fields. All day long they gorged themselves on the insects until the pests were vanquished and the people were saved. The heaven-sent birds then returned to the lake islands."

Unfortunately, other settlers brought pest insects and weeds with them and there were no predators to destroy them. Often these pests came to America by ship from Europe or Asia in cargoes of seeds, wood, and other needed products. Benjamin Walsh, an entomologist, wrote about these pests shortly after the Civil War:

"It is a remarkable fact that fully one-half of our worst

Certain crickets cause a great deal of damage to farmers' crops.

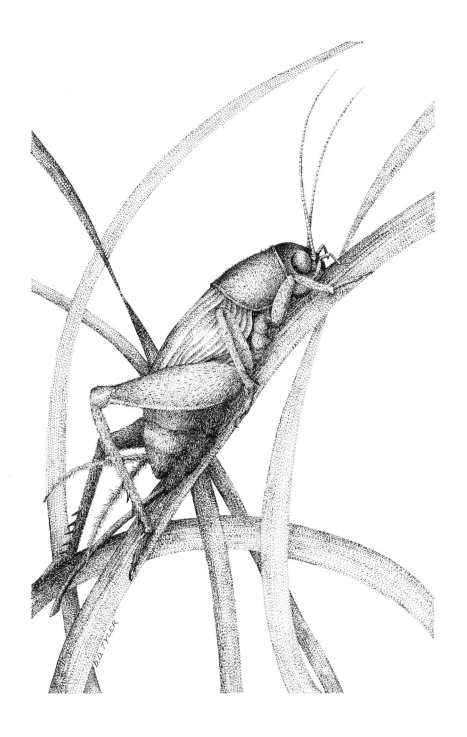

D.D. TYLER

insect foes are not native American citizens, but have been introduced here from Europe. The plain common sense remedy for such a state of things is, by artificial means, to import the European parasites that in their own country prey upon the wheat midge, the Hessian fly, and other imported insects that afflict the North American farmer. Accident has furnished us with the trouble; science must furnish us with the remedy."

A few years after Benjamin Walsh wrote those words, a terrible pest invaded California from Australia. It was the cottony-cushion scale, a very small, whitish insect that attaches itself in colonies to trees and sucks their juices. No one was quite sure how the insects had arrived, but soon they spread through California's orange groves, covering the trees like snow.

The scale insects threatened to destroy the new California orange industry. "The white scales were incrusting our orange trees with a hideous leprosy," a scientist wrote afterward. "They spread with wonderful rapidity and would have made citrus growth on the whole North American continent impossible within a few years."

The orange growers turned to the USDA for help. Money was raised and a "bug hunter" named Albert Koebele was sent to Australia in 1888 to find out about the cottony-cushion scale. Koebele searched through a large area of Australia but could find very few of those insects. He realized that something was keeping the cottony-cushion scale from becoming a pest in its native land.

With the help of Australian entomologists, Koebele finally found the cottony-cushion scale in a few places. He also found several of its natural enemies feeding on it—a parasitic fly, a green lacewing, and a ladybird beetle. The ladybird, whose scientific name is Vedalia, ate enormous numbers of the scale insects.

Koebele collected a number of these Vedalia beetles and packed them in wooden boxes with sufficient scale insects to feed on. Then he took them aboard ship, put the boxes on ice, and sailed back to California. He and a colleague released them in several orange groves near Los Angeles early in 1889.

Albert Koebele and the Vedalia beetle soon became famous among entomologists and agricultural experts all around the world. The beetles multiplied rapidly, spread from tree to tree, and began to destroy the cottony-cushion scale. The orange growers were astonished and very much excited. One of them wrote:

"The Vedalia has multiplied in numbers and spread so rapidly that every one of my 3,200 orchard trees is literally swarming with them. All of my ornamental trees, shrubs, and vines which were infested with white scale are practically cleansed by this wonderful beetle. People

Vedalia beetles feeding on cottony-cushion scales.

are coming here daily, and by placing infested branches upon the ground beneath my trees for two hours can secure colonies of thousands of the Vedalia. Over 50,000 have been taken away to other orchards during the past week, and there are millions still remaining."

By the end of that summer the cottony-cushion scale was no longer a pest. It was a wonderful achievement, with humanity using science to correct a problem brought on by accident. The future of pest control looked promising. But later on, other people forgot science and much greater trouble lay ahead.

4
THE DREADED GYPSIES

The arrival of the gypsy moth in the United States was not an accident. Shortly after the Civil War a French naturalist named Leopold Trouvelot, who was living in Medford, Massachusetts, conducted experiments with several kinds of insects to see if they could produce silk. His experiments produced not silk but an ecological disaster.

Silk was an important product at that time, before synthetic fabrics such as rayon and nylon, which are made from chemicals, came into use. It was in great demand for the manufacture of fine dresses, shirts, curtains, and other common articles. Most raw silk comes from the mulberry silkworm, a moth which spins a fine yellowish fiber in making its cocoon. Trouvelot hoped to raise other moths in Massachusetts whose cocoons could be unwound for their silk.

The Dreaded Gypsies

The gypsy moth is a distant relative of the mulberry silkworm. It is a very hardy insect and lives in a great variety of climates extending across Europe, North Africa, and Asia, to Siberia, China, and Japan. It is not native to either North or South America. But Trouvelot guessed that this hardy insect would do well in the New England climate.

Trouvelot imported some eggs of the gypsy moth from Europe. When the eggs hatched he put the caterpillars on shrubs in his backyard where they could feed on the leaves. To keep them from escaping, he covered them tightly with nets.

Trouvelot's gypsy moths did well in captivity, but nothing ever came of his experiments because gypsy moths do not spin cocoons of silk. His Massachusetts neighbors, however, were in for an unpleasant experience.

The disaster began as it usually does in horror movies —with heavy rain, thunder, and lightning. The storm that struck Medford that early summer evening in 1869 tore the nets off Trouvelot's shrubs. The gypsy moth caterpillars escaped. They found the trees around Medford to their liking. Slowly they spread through the town. Within a few years they became one of the most serious insect pests New England has ever had.

What is this awesome creature which was to bring so much grief to Leopold Trouvelot's neighbors? It appears in the woods in late spring as a little brownish-yellow caterpillar bristling with pale yellow hairs, chewing its way out of its egg to freedom. Then, spinning a long silken thread, it suspends itself from branches and is often carried long distances by the wind. It has been recorded that one gypsy moth caterpillar, just out of the egg, spun a thread that was sixty-nine feet, four inches long.

The young caterpillar grows rapidly, feeding at night on a great variety of leaves, especially those of oak, birch, poplar, and fruit trees. In daylight it crawls away to hide in crevices of bark or among leaves on the ground.

By early July the caterpillar is about two inches long. Some people (though they are hard to find) say it is a handsome beast. It has a large head. Its back is dotted with blue and red spots, and tufts of pale yellow hairs sprout from its sides. At this stage the caterpillar spins its cocoon and then emerges several weeks later as a moth.

As moths, the males and females don't look very much alike. The male is colored a ruddy brown. In fact, the moth got its name in England, where the people said it reminded them of a gypsy's healthy complexion.

The female moth is much larger, about an inch long.

A male gypsy moth (above) and a gypsy moth larva feeding on an apple leaf (below).

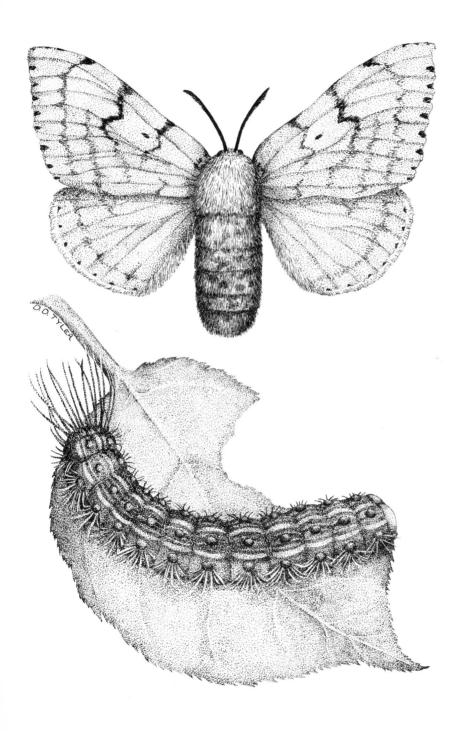

Its broad white wings are marked with black zigzag lines. Unlike the male, the female cannot fly because its heavy body is packed full of eggs. When it comes out of the cocoon, it simply crawls a few inches away and attracts a male with its scent.

The scent is easily picked up by males some distance away and this may prove to be the gypsy moth's undoing. After mating, the female lays a cluster of eggs in a protected place such as the underside of a stone or a branch. There may be five hundred eggs or sometimes even a thousand in a cluster. The female moth spins a covering of brownish-orange fuzz over the eggs and then dies.

This was the creature that began to spread out over eastern Massachusetts. Some of the people who lived there told scientists later on what it was like. One woman recalled:

"We moved to Medford in 1882. The caterpillars were over everything in our yard and stripped all our fruit trees, taking the apple trees first and then the pears. There was a beautiful maple on the street in front of the next house, and all its leaves were eaten. They got from the ground up on the house and blackened the front of it. They would get into the house in spite of every precaution and we would even find them on the clothing hanging in the closets."

By the 1880s the gypsy moths were all over town. Another woman said:

The Dreaded Gypsies

"My sister cried out one day, 'the caterpillars are marching up the street.' I went to the front door, and sure enough, the street was black with them, coming across from my neighbors and heading straight for my yard. They had eaten her trees, but our trees at that time were only partially eaten."

People gathered the caterpillars in baskets and burned them. When they went out for a walk, they felt the insects fall from trees onto their hats or down their collars. Some residents carried umbrellas to protect themselves. Women returned home with their long skirts stained by the caterpillars they had squashed underfoot. One man reported that the caterpillars were so thick on the trees that they were stuck together "like cold macaroni."

As the gypsy moth spread and property owners complained, the state government decided to help. Crews were hired to find the egg masses and paint them with creosote. They sprayed the caterpillars with arsenate of lead, an old-fashioned insecticide.

One of the most successful methods they used in parks and around homes was to tie a burlap bag, its flap hanging down, around the trunk of a large tree. At daybreak, when the caterpillars started down the trunk to find hiding places, they crawled under the burlap bag. There they were easily collected and burned.

For a while, the gypsy moths heavily damaged the forests of eastern Massachusetts. Scientists in both the state and federal governments studied these insects and realized what the problem was. They knew that in Europe and Asia the gypsy moth population rapidly increases from time to time and damages orchards and small forests. But these outbreaks usually do not last very long. In its native habitat there are many predators and parasites that feed on the gypsy moth and keep it under control.

When Leopold Trouvelot's gypsy moths escaped, they found no such predators and parasites waiting for them. They had few natural enemies in the New World. The United States government found out what kinds of predators and parasites attacked them in their natural habitat, and imported more than forty of the gypsy moths' natural enemies from Europe and Japan. At least eleven of them learned to live in the New England climate. One insect from Japan, called a tachinid fly, was especially successful in attacking gypsy moths in the United States.

As time went by, many American species learned to feed on gypsy moths too. Chickadees and other birds searched along tree trunks and branches for the eggs, cocoons, and caterpillars. An animal that took a special liking for the cocoons of gypsy moths was the white-footed mouse. This small rodent, which is numerous in New England, ate large quantities of the cocoons before they had a chance to develop into troublesome caterpillars.

The white-footed mouse feeds hungrily on a pupa of the gypsy moth.

Bug Hunters

Although the gypsy moth spread from Massachusetts into neighboring states, it lost some of its reputation as the terror of the forest. Here and there it would have a population explosion and damage part of the forest. But then nature would step in. The gypsy moth population would collapse again as starvation, disease, parasites, and predators got the upper hand.

It seemed that the dreaded gypsies had been tamed by nature itself, but often people do not know how to let well enough alone. Experts in insect control suddenly found a new weapon to get rid of gypsy moths and other pest insects. But the new weapon backfired.

5
A DEADLY RAIN

Just before World War II a Swiss scientist named Paul
Müller made some experiments to find a chemical that
would keep moths out of clothing. He concocted a powdery
substance that proved to be very effective. The smallest
amount killed all sorts of insects. He was delighted to
find that the chemical went on killing insects for many
months after he applied it to any surface.

The new chemical was called DDT. During World War
II it was used by the U.S. Army and other military units
to wipe out mosquitoes that caused malaria, yellow fever,
and other diseases. As a powder, DDT was dusted on
human beings to kill disease-carrying lice. Müller was
awarded a Nobel Prize in physiology and medicine. Many
people called DDT, which was cheap and easy to manufacture, a "miracle" chemical that would solve the problem of pest insects.

After the war, the new pesticide became available to the public. Whenever farmers or foresters had problems with insects, they immediately sprayed them with DDT. A few scientists warned the public that it was a poison and it should be used with great care. But their voices were unheard in the general excitement about DDT's ability to kill pests.

During the 1950s the USDA decided that the time had come to get rid of insect pests once and for all. One of the first targets was the gypsy moth. The USDA began a program to eradicate—or wipe out—the moths in the United States.

Low-flying planes sprayed one million acres of the forests in the Northeast with DDT. During the second year of the program, the USDA and state agricultural agencies sprayed three million acres.

One of the chief target areas was Long Island in New York State, where gypsy moths had spread. The USDA justified spraying that region of farms and suburban homes by saying moths might infest New York City itself. This was silly, because the city has comparatively few trees and they are not the kinds that gypsy moths prefer.

The USDA's planes spread their rain of poison from the skies. Some gypsy moths were destroyed, but so were songbirds, fish, crabs, and helpful insects. Beekeepers lost their entire colonies.

A Deadly Rain

"It is a sad thing," a beekeeper said, "to walk into a yard in May and not hear a bee buzz."

Many people were alarmed because even milk became contaminated by DDT throughout parts of the Northeast. Cows grazed where the spray fell and stored the poison in their bodies. The USDA and some state agencies did not listen to complaints from people who were worried about this great campaign to spread a poisonous chemical on millions of acres of fields and forests. These agencies had abandoned the scientific approach to pest control. It was a period of wars around the world, and the agencies used the methods and the language of war.

"We must *wipe out* insects," they were saying. "We must *win the war* against insects."

Expensive programs were planned to destroy insects, using DDT and other new chemical pesticides very much like it:

• The USDA began an all-out war against fire ants in the Southeastern states. These insects had been accidentally imported from Argentina some years before, but the states did not consider them to be one of their most serious pests. Fire ants do not eat crops, but feed on other insects. While their bite is painful, they are no more of a problem than wasps and bees in that respect.

Yet the USDA spent millions of dollars to spread the new pesticides over millions of acres over the country. Wildlife experts protested when thousands of quail, wild

Chemical pesticides such as DDT were used in great quantities to kill the cotton boll weevil.

turkeys, woodcock, and other game birds were killed by the effects of the poisons. Farmers protested when their cattle, horses, hogs, and chickens either died or produced dead young.

• Although the Vedalia ladybird beetle had kept the cottony-cushion scale under control on California orange trees for more than fifty years, agricultural experts decided to spray the trees to get rid of other pests. DDT killed the Vedalia beetles, and within a few months cottony-cushion scale insects again covered the orange trees. Many trees died. The growers began to pay a dollar apiece for Vedalia beetles to get them back on their trees. They stopped using DDT, or used it much more sparingly, and the Vedalia beetles once more controlled the scale insects.

• Foresters in the Northeast declared war on the spruce budworm, an insect which attacks evergreens such as spruce and fir. They sprayed millions of acres with DDT. The chief result of the program was that great numbers of salmon and trout, as well as the harmless insects they feed on, died in streams and rivers that flow through the forests.

• At California's Clear Lake, north of San Francisco, vacationers were bothered by gnats that bred in the water. The lake was sprayed with very low quantities of a close chemical relative of DDT. Still, most of the lake's water birds, including western grebes—which are also called "swan grebes"—died or were not able to hatch their young. And fishermen were taking home contaminated fish.

When scientists investigated they discovered a frightening thing about DDT and its chemical relatives. Although very small quantities had been used, the chemical was "magnified" by the food chain in the lake. Small plants called plankton absorbed tiny quantities of the chemical. Plant-eating fish took in the chemical from the plants. When larger fish ate the smaller ones, they stored large quantities of the chemical in their bodies. The poison, in turn, was passed on to the fish-eating birds. When the grebes and other birds stored up enough of the poison in their own bodies, they died.

What advantage came from these widespread and costly spraying programs? Very little. The hardy gypsy moth was not eradicated, as the USDA had promised, but it actually began to increase its numbers and spread into areas where it had not lived before. DDT had done more damage to the parasites and predators. During the year following the spray, the gypsy moth eggs hatched and the moth population built up again easily because most of the birds and insects that fed on it were no longer there. The story was the same with the USDA's other programs. The fire ant went on spreading through the Southeast. The orange growers had nearly destroyed the most successful predator insect of them all—the Vedalia beetle. The spruce budworm went on spreading through Northeastern forests. The state of California stopped the spraying at Clear Lake because fishermen were taking home contaminated fish.

This all-out war against insects might have gone on for many years except for a courageous scientist named Rachel Carson. Like many other people she became worried about the widespread destruction caused by pesticides. Fortunately, she was a gifted writer who had written best-selling books about the sea and its life. In 1962 she wrote a book called *Silent Spring*, which described the foolish and expensive chemical warfare against insects.

Much of the material for the book was given to Rachel Carson by scientists in the USDA and other government agencies. They, too, did not like the unscientific methods those agencies used to control insects, but they would have lost their jobs if they had spoken out in public.

Rachel Carson believed that the use of DDT and its chemical relatives should be forbidden because they did not break down in the environment. They remained in the soil or in the tissues of animals and were passed along from one creature to another. But she did not write that all pesticides should be discarded. She believed that pesticides, if they broke down quickly in the environment,

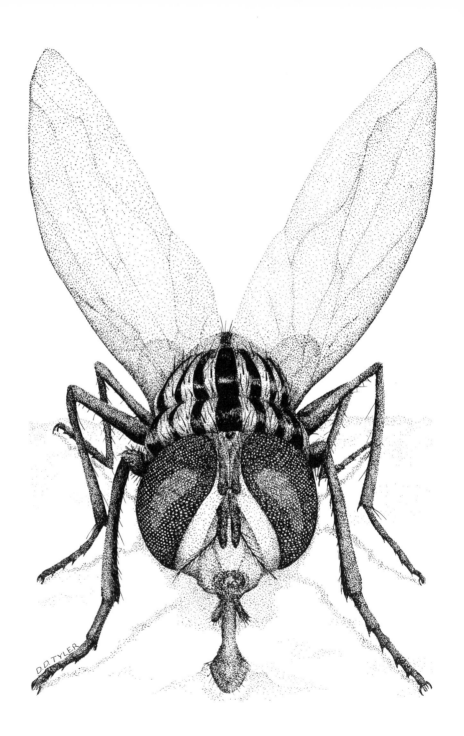

could be used sometimes when pest insects were not controlled by nature.

Although *Silent Spring* became a best-seller, the chemical companies and their friends in government attacked it fiercely. Rachel Carson had written things they didn't want to read. She showed that the government no longer practiced biological control against pest insects. Farmers used large quantities of poisons to kill pest insects, but they killed their parasites and predators too. So the pests returned every year in greater numbers and farmers had to spend more and more money to buy pesticides. The only people to benefit from this all-out war against insects were the chemical manufacturers.

"It is not my contention that chemical insecticides must never be used," Rachel Carson wrote in *Silent Spring*. "I do contend that we have put poisonous and biologically potent chemicals into the hands of persons largely or wholly ignorant of their potentials for harm. I contend, furthermore, that we have allowed these chemicals to be used with little or no advance investigation of their effect on soil, water, wildlife, and man himself. Future generations are unlikely to condone our lack of prudent concern

---

Pest control experts thought they could get rid of the house fly with DDT. They were wrong.

for the integrity of the natural world that supports all life."

President John F. Kennedy read Rachel Carson's book and ordered government agencies to take a new look at the way they attacked pest insects. Other scientists, inspired by Rachel Carson's words, turned up new evidence against DDT. The bald eagle, our national bird, and a number of other creatures were threatened with extinction because of pesticides. Many scientists, noticing the increase of cancer in America, said that many cancers could be traced to chemicals in our environment.

Rachel Carson died of cancer in 1964, but her friends carried on her struggle. The National Audubon Society and other conservation organizations joined scientists in demanding that the government forbid the use of DDT. The chemical industry fought hard to keep it, but scientific evidence was against it. In 1973 the federal government banned most uses of DDT in this country.

Rachel Carson had awakened the country to the danger of relying only on deadly poisons to solve our pest insect problems. These poisons do not select their target. The "pest" in the word "pesticide" had fooled people. They thought the chemical killed only "pests." But our war against pest insects had become a war against life itself.

6
TURNING BACK TO SCIENCE

Someone has said, "A weed is simply a flower that is growing where people don't want it to grow." A farmer may think that the milkweed in his field is a pest—it takes up space that he would rather see used by plants that his cattle like to eat. Another person, who likes to look at wild flowers and the butterflies they attract, might think the milkweed is a beautiful and important part of the field.

The same thing can be said about insects. Some people believe that all insects are pests. "The only good bug is a dead bug," they say. Wiser people know that most insects are not pests at all. Insects have a role to play in the natural world, and only a very few of them grow numerous enough in some places to become a special problem.

As Benjamin Walsh said over a century ago when he was talking about pest insects, "Science must furnish us with the remedy." Unfortunately, the pest control experts of our own time forgot that advice. Then Rachel Carson and a few other scientists began to advise them to take another look at the possibilities of biological control.

In several parts of the Northeast, people were taking this advice. Gypsy moths invaded an oak forest in New Jersey's Morristown National Historical Park. Several government agencies wanted to spray the park with Sevin, a pesticide that many experts say is not as harmful to the environment as DDT.

But the National Park Service, which runs the park, refused. Scientists had recently released a few parasites of the gypsy moth there, and it was decided to give them a chance to work. Sevin would probably have killed most of the parasites.

For a while it looked as if the National Park Service had made a mistake. The gypsy moths kept spreading into the park. By eating the foliage, these insects killed more than ten thousand oak trees and damaged many more.

Suddenly, in the third year of their invasion, the gypsy moths nearly disappeared. When a population of insects builds up to large numbers, they are often attacked by a virus or other disease. A virus struck the gypsy moths in the Morristown park, and most of them died from the disease. The parasites and natural predators finished them off. The gypsy moth population in the park had collapsed completely.

"It is practically impossible to find any significant number of egg masses here now," the superintendent of the park said a year later.

The superintendent also said that the forest was healthier than it had been before the gypsy moths arrived. They had thinned out the forest, leaving more room for the other trees, while three-quarters of the trees remained in excellent health. Young trees other than oak were coming in—a forest is healthier and better able to resist diseases and pest insects if it is composed of several different species.

The trouble was that in many areas of the country there were no parasites or predators to control insects when they became pests. Pesticides, even the supposedly safer ones that replaced DDT, had wiped them out. Yet the pest insects themselves in many cases were becoming resistant to pesticides. The chemicals did not kill them as easily as they did at first.

Scientists first began to notice this resistance in the case of DDT and houseflies. The housefly often becomes a pest around houses and farms. At first it seemed that DDT would solve all our problems with houseflies. But these insects developed a resistance to pesticides in the same way that many other insect pests have done since that time.

How does an insect become resistant—or immune—to pesticides? When DDT was used against houseflies, most of them died. But in every population of animals, scientists find a few individuals whose genes are slightly different from those of the others. Some houseflies proved to be able to withstand DDT better than others. After the spraying, they were the only ones left alive. When they reproduced, they passed on their characteristics to their young. Within a short time the housefly population was resistant to DDT.

"The pest control experts thought that all they needed was pesticides," an entomologist with the U.S. Forest Service says. "So when they started using DDT they did very little new research on biological controls. Then, when they found out that chemicals did not really solve the problem of pests, they weren't ready to switch to biological control."

But here and there, a few scientists were at work trying to get nature's help once more. Rachel Carson particularly praised the work of A. D. Pickett in Nova Scotia. For many years Pickett searched for ways to control codling moths, which damaged the apple crop in his part of Canada. He noticed that when DDT was used to kill codling moths, it also killed the predators that used to control mites on the trees. The mites then nearly destroyed the apple crop.

"We move from crisis to crisis, merely trading one problem for another," Pickett said.

The codling moth larva is a serious pest on apple trees.

Pickett began to use pesticides sparingly. When he was forced to use pesticides, he timed his spraying so that it did not kill the predator insects that were still in the egg stage. He looked for chemicals that would do the least harm to other kinds of insects and he used them in very small amounts. The apple growers who followed his advice produced fruit of high quality while spending only about one-tenth as much as usual on chemicals.

Some of the most interesting work in biological control in recent years has been done at Michigan State University. Brian Croft, a scientist there, began to study the spider mite. This mite damages fruit trees, and growers have spent many thousands of dollars trying to control it with chemicals. After a while the spider mite became resistant to pesticides.

Brian Croft knew that there are several different kinds of mites on fruit trees, but not all of them are pests. One of these mites is a predator that attacks spider mites. For many complicated reasons which even scientists do not fully understand, the plant-eating insects that become pests often develop resistance to pesticides, but their predators and parasites rarely become resistant to chemicals.

But the mite that eats spider mites also became resistant. Croft took advantage of this fact, with great skill. The predator mite spends the winter in the grass

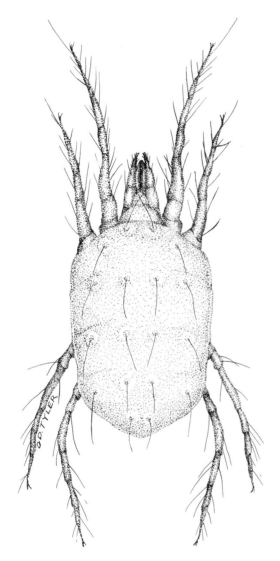

Mites are not insects. They are tiny relatives of the spider. Scientists use predatory mites to fight pest mites on apple trees.

under the apple trees. But farmers usually clear out the grass under their fruit trees to conserve soil and moisture and discourage mice that sometimes also live there. Mice often damage fruit trees too.

Croft discovered a way to vary the vegetation under the trees. Sometimes he grew grass there and at other times weeds. The constant change upset the mice and they left. But the predator mites remained.

"Now the predator mites eat the spider mites that aren't killed by pesticides," Brian Croft says. "We've even taken these predator mites that are resistant to pesticides and put them in other parts of the country. Then they interbreed with local predator mites and form a new population that is resistant to certain chemicals."

The pest control experts were turning back to science.

7
THE SEARCH

The cereal leaf beetle that is now being used in Indiana became the object of one of the most successful struggles that Americans have carried on against an insect pest in recent years. Curiously, the farmers whose crops were attacked by this beetle were not aware for several years that it even existed.

Southern Michigan is an important region for growing grain crops such as oats and barley. In 1959, farmers in the area noticed that unknown insects were attacking the leaves of their grain plants. During the next two or three years agricultural experts in the federal and state agencies began to receive a stream of complaints from grain growers. No one was sure what damaging new insect was at work in the grainfields.

Finally, in 1962 some of the pests were collected in a

Michigan field. Experts identified the culprit as the cereal leaf beetle. Like so many of our serious insect pests, the cereal leaf beetle is not native to the United States. It lives in many parts of Europe and across the Soviet Union all the way to Siberia.

How did the cereal leaf beetle reach Michigan from Europe or Asia? Obviously, this little creature did not fly thousands of miles to land in a Midwestern grainfield. But throughout American history various insects have reached our shores hidden away in plants, lumber, seeds, grain, and other products brought by ship from abroad. The USDA operates quarantine stations at our ports of entry to inspect shipments for pest insects. When inspectors find such pests, they either fumigate the cargo—disinfect it with smoke, gas, or some other vapor—or refuse to let it enter the country.

The USDA inspectors discovered cereal leaf beetles in shipments from abroad at least six times between 1957 and 1962. But once during that time it must have slipped in unseen. Some experts call this beetle a "jet age insect" because it is the first serious insect pest to come *directly* to the Midwest from the Old World without having been discovered first near the coast. The cereal leaf beetle may have arrived in the Great Lakes on a ship entering through the St. Lawrence Seaway, or even on a cargo carried by jet to Chicago or Detroit.

The Search

In Europe, the cereal leaf beetle feeds on a number of grasses as well as on such small grains as wheat, oats, and barley. As it spread swiftly through the Midwestern United States, it also attacked corn. In this age of world-wide food shortages, an insect that threatens grain crops —the staple of bread, cereals, and livestock feed—is cause for worry.

Farmers, with the help of the USDA, tried to wipe out the cereal leaf beetle before it gained a foothold in the United States. They used planes to spray grain fields with chemical pesticides. They set up Midwestern quarantine stations to inspect all shipments of crops or hay to keep the beetles in the area where they were first discovered. But nothing stopped the beetles' spread.

"These are flying insects," a USDA entomologist says. "They'd get up in the air and the wind would carry them as far as thirty miles—over the quarantine stations and out of reach of the spray planes. We sent up small planes with nets attached to them and caught cereal leaf beetles fifteen hundred feet above the ground."

Helped by the wind, cereal leaf beetles fanned out from Michigan into Illinois, Indiana, Kentucky, Maryland, Massachusetts, Missouri, New Jersey, New York, Ohio, Pennsylvania, Tennessee, Virginia, West Virginia, and Wisconsin.

"They haven't spread much west of the Mississippi

River," the USDA entomologist says. "The winds haven't been right, I guess, and the beetles seem to like smaller fields with lots of bushes nearby for shelter, rather than the wide open spaces of the West. So far they don't seem to like the warmer weather in the Deep South either. We've got our fingers crossed."

But entomologists realized they needed something more than crossed fingers to save grain growers in the Midwest and Northeast from the cereal leaf beetle. Researchers at the USDA began to study this insect closely, hoping to find its weak points. They knew they had to work fast. The great surpluses of grain stored in countries such as the United States and Canada were disappearing as drought and famine struck other parts of the world. Millions of people were going hungry. The world could not afford to lose an important source of grain to a small insect.

The USDA was equipped to make a thorough study of the cereal leaf beetle. For many years its Agricultural Research Service has kept a laboratory outside Paris, France. Entomologists there make studies on pest insects

---

The cereal leaf beetle and its eggs on oat leaves. Because the larvae do not eat all the way through the leaves, the white inner skin remains.

and other insects that feed on them. They were able to put together a description of the cereal leaf beetle's life cycle.

The adult beetles spend the winter hibernating in the grain stubble or in the bushes or thickets that surround the fields. As the days grow warmer in early spring, the beetles crawl from their hiding places and search for the tender young grasses and grain plants. There they begin to feed and mate. The females lay their eggs on the leaves.

The larvae hatch while the leaves are still tender and begin feeding on them too. Later they drop to the ground, where they burrow into the soil, make little cells of earth to cover themselves, and pupate. Pupation is the stage in an insect's life when the larva changes into an adult. The new generation of adults comes from the cocoon or cell in late June and July. They feed on grain leaves until the cool days of autumn drive them into hibernation.

In Europe, the USDA's entomologists carry on research on both sides of the Iron Curtain. When they began to study the cereal leaf beetle's habits, they learned that it could be found throughout most of Europe but it was seldom numerous enough to cause damage. In some years these beetles become pests in small areas of southeastern Europe, including the Soviet Union, Yugoslavia, and Rumania. But while these insects also live in countries such as Italy, France, Denmark, Sweden, and Spain, they do little damage to crops.

The Search

The cereal leaf beetle, then, was not a serious pest in its natural habitat. It fitted well into its environment. When its numbers increased for a while, nature usually stepped in and brought its population back to normal. Like most other animals, it had natural enemies that kept it from ever becoming too numerous.

The cereal leaf beetle had no such natural enemies in the United States. It just kept on increasing its population until it became a pest. The job that lay before the USDA's scientists was to find those natural enemies and use them to control the cereal leaf beetle in its new home.

8
WILY WASPS

The USDA's scientists had to work fast. The cereal leaf beetle was spreading into new states each year. Should the scientists look for predators or parasites to attack the beetles?

They decided to look for parasites. Although predatory insects are valuable enemies against some pests, biological control experts today usually prefer parasites because they are "host specific." That means they feed mainly on the pests they have been introduced to attack, and are not tempted by every other kind of insect that passes by.

The experts at the laboratory in France needed to find out what kind of parasites attack the cereal leaf beetle. They visited meadows and grainfields all over Europe, searching for the eggs and larvae of this pest insect. The biological control experts found the beetles in such places as Saragossa, Rome, Cahors, Seville, Villefranche, Kiev,

Wily Wasps

Lidköping, Ystad, Ybbs, Huntley, Frederikshavn, Lentilly, and many more.

The entomologists picked the beetles' eggs and larvae by hand from grasses and grain plants and shipped them to the USDA laboratory outside Paris. In the lab, other entomologists dissected—or cut open—eggs and larvae to learn what was inside them. They found at least eight different kinds of parasitic insects feeding inside either the eggs or larvae.

The next step was to decide which of these parasites were worth sending to the United States. Several of the parasites were discovered in only a few larvae, so they did not seem important enough to bother with. A certain kind of fly was found in many larvae, but the entomologists decided not to study it further. They had tried to introduce this fly into the United States some years before to control a beetle that was a pest on asparagus plants, but it had died out quickly in its new surroundings. They needed a hardier parasitic insect that would be able to live in the different ranges of temperature and humidity found in Michigan, West Virginia, and New Jersey.

Luckily, they discovered four different kinds of parasites that seemed to be good candidates for life in the United States. Each of them in Europe feeds mainly on the eggs or larvae of the cereal leaf beetle. Each of these four parasites was a tiny wasp.

Wasps have a bad reputation. When we hear about wasps, we are likely to imagine a swarm of large, angry insects heading in our direction, ready to jab their long stingers into our tender skin. Some wasps, such as yellow jackets or the larger kinds that we call hornets, are able to deliver a painful sting when we bother them. But wasps generally will not hurt us if we leave them alone. They are among the most interesting animals in the world, and many of them are helpful to people.

Wasps belong to an order of insects called Hymenoptera, which means membrane-winged. This order includes the wasps, bees, and ants. Like most kinds of insects, they have four stages in their life cycle. They begin life in the egg. The young or larvae hatch from eggs. After feeding and growing for a while, most species weave a cocoon and then change into a pupa. Finally, they come out of the cocoon as mature insects.

Some kinds of wasps are social insects. They build nests and live together in a colony that has many members. The paper wasps, for instance, build their nests from paper, which they make from chewing small strips of wood and mixing it with their saliva. In fact, European scientists are said to have gotten the idea of manufacturing paper from wood pulp by watching paper wasps.

Other kinds of wasps live alone rather than in colonies. The mud dauber is a familiar solitary wasp, which plasters its nest of mud and saliva in protected places. The female mud dauber hunts spiders or sometimes caterpillars, which she paralyzes with her stinger and brings back to the nest to feed her young. Because she nests during the warm months, the flesh of the spider or caterpillar would rot in the heat before her tiny young finished feeding on it. By paralyzing instead of killing the unlucky prey, she makes sure that it remains alive and fresh for her young.

There are also very close relatives of the wasps that have found a different way to feed their young. These insects are often called parasitic wasps. There are many different kinds and most of them are very small—about the size of a gnat. Their stinger is not used for stinging other creatures but for laying eggs. It is called an ovipositor.

These tiny wasps do not build nests. Instead of killing or paralyzing other insects and carrying them back as food for their young, they use the ovipositor to plant their eggs directly on or into the eggs or bodies of their hosts.

Now another decision had to be made. Should the USDA choose one of these parasitic wasps to release in the United States, or send all four of them? Again, the lack of time helped the entomologists to make up their minds. If only one parasite was chosen and for some reason it did not succeed in its new home, several years would have been wasted. The USDA's experts decided to try to establish all four of them.

To make certain there was a collection of parasites from many different climates, the USDA's experts gathered them from all over Europe for shipment to the United States. Some of the wasps were already adults, having eaten their way out of their hosts. The entomologists captured them in glass tubes, then "blew" them into food cartons. Later they cut openings in the cartons and covered them with plastic screen. They put honey and water on the screen to feed the adults.

The shipments reached the United States within eight to ten hours after they were delivered to the airport at Paris. Other parasites arrived as larvae inside the eggs or larvae of cereal leaf beetles. The shipments were examined at the USDA's laboratory in New Jersey.

"We looked them over to make sure the parasites were healthy," an entomologist at the lab says. "We also checked them to make sure they hadn't been parasitized themselves. After all, we didn't want any parasites on our parasites!"

Then the parasites went on to the Midwest. The USDA

A tiny parasitic wasp called Anaphes *lays its egg in the egg of the cereal leaf beetle. When a wasp larva hatches, it eats the beetle's egg. This illustration is greatly enlarged.*

sent some to agricultural scientists at Michigan State and Purdue universities, which were cooperating in the program. But the USDA also built a special laboratory in 1966 at Niles, Michigan, which became the headquarters of the program against the cereal leaf beetle.

The new headquarters was named the Cereal Leaf Beetle Parasite Rearing Laboratory. Director Tom Burger points out that the insects studied by his staff are not really parasites—and, in fact, they are not true wasps.

"The news that we were going to spread wasps over the countryside scared a lot of people," Burger says. "They were thinking of bad-tempered hornets, I guess. But, you know, these insects are not considered true wasps because they don't really sting. They're very close relatives, though, and so they are called parasitic wasps.

"Other people got scared because they heard we were going to release a lot of parasites. They were thinking of ticks or tapeworms. We even made it worse by using 'parasite' in our lab's name. But these insects aren't true parasites either. True parasites, like tapeworms, feed on their host, but don't usually kill it. These insects kill their host eventually, so some scientists prefer to call them parasitoids—which means *something like* a parasite."

But sometimes even scientists don't have the last word. Anyone who refers to these wasplike, parasitelike insects as either parasites or wasps is very close to being right.

9
THE LAB

Vera Montgomery heard about the new laboratory when it opened in Niles, only a few miles across the state line from her home in Indiana. Because she had always been interested in plants and animals, she was curious about what was going on in the lab, but it seemed very remote from her daily life. She and her husband had worked hard most of their lives. They had raised eight children, several of whom had already left home to go to college. Vera Montgomery thought of herself as a housewife with little time to devote to an outside career.

She had grown up on a farm. Her father taught her a great deal about the natural world and gave her the skills to grow plants of her own. She spent many hours wandering around the farm, looking at the plants and animals, learning their names and their habits.

Vera Montgomery did not forget those days after she

married. When her own children were old enough, she took them on walks and showed them the plants and animals she had learned about as a girl. Sometimes her children asked her questions about nature that she could not answer. Then she would go to the public library and take out books on natural history.

"When my son Steve was a young boy he said to me one day, 'I can't wait for the butterfly weed to bloom,'" Vera Montgomery remembers. "You see, he knew that when those plants bloomed many interesting insects would come to them. My children learned so quickly about nature that it was always an adventure to go into the fields with them."

Vera Montgomery and her children often talked about the new insect lab at Niles. They wondered what sort of work the entomologists did with beetles and their parasites. Then one day two of her sons came home with some interesting news.

"Guess where the woman across the street got a job?" one of them called.

Experts in biological control study all kinds of insects to find suitable parasites and predators on pests. The green lacewing (above) is a delicate-looking insect, but its larva (below) is a fierce predator on pests such as aphids and is often called the "aphid lion."

of beetles' eggs and larvae in the laboratory—the staff had no trouble finding them in nearby fields—and simply let the wasps parasitize them. In that case, the staff could control the process every step of the way, and later release the new generations of wasps as they appeared.

But these parasitic wasps are, after all, wild animals. Little was known about them at the time. No one saw all the difficulties that would appear once the program got under way.

Vera Montgomery's first job was picking beetles' eggs off oat leaves. She worked fast, plucking off the small yellow eggs with dissecting needles and setting them on glass plates. Whenever she had a few minutes to herself, she watched the process by which the tiny parasitic wasp, *Anaphes*, attacked the beetles' eggs:

"We put the beetles' eggs into little glass cases, and in another case we had eggs that had been parasitized by *Anaphes*. The *Anaphes* spends most of its life cycle in the egg of another species of insect—first as an egg itself, then as a larva, feeding on the beetle's egg, and then as a pupa. When it becomes an adult it finally emerges from the beetle's egg."

She watched *Anaphes* as they emerged, and started their adult lives.

"I learned to tell the males from the females. The females have knobs on their antennae, and the males do not.

The male *Anaphes* stands around and waits for a female to come from the beetle's egg and then they mate. A female *Anaphes* can produce young even if she doesn't mate, but then the young would only be males. To get female young, the mother must mate first."

Workers in the lab put the newly mated females into the glass cases holding the eggs of the cereal leaf beetle. The *Anaphes* female flies from egg to egg, visiting at least five or six of them and depositing several eggs in

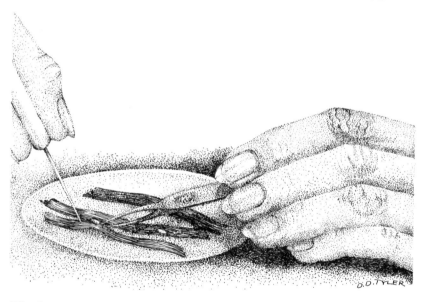

Workers in the laboratory need sharp eyesight and agile fingers to remove cereal leaf beetle eggs from oat leaves.

each of them. When those beetles' eggs, carrying their cargo of parasites, were distributed in grainfields, they would never hatch. Instead, after a few weeks the adult wasps would emerge and search for other beetles' eggs. In that way, *Anaphes* soon became a wild resident of Midwestern grainfields, living its life there and destroying cereal leaf beetles.

The lab's staff members ran into many difficulties. Although *Anaphes* did well in captivity, the cereal leaf beetle itself became a problem. During a part of its life cycle, the beetle enters a state which entomologists call diapause—a resting period when it stops all growth and activity. The staff members could do nothing but pack the beetles away in cold storage for many weeks until they became active again.

"Many of them die in cold storage," Vera Montgomery says. "And even those that live take about five months to raise, which is much too long for our kind of program. We had to turn out eggs fast so that we could rear a lot of *Anaphes* too."

The lab's staff members had to invent new ways to carry out their program for rearing parasites. Vera Montgomery was given a chance to help on a scientific project. In some of the beetles' eggs brought in from grainfields, the staff members found another parasitic wasp called *Trichogramma*.

"I was asked to rear this parasite in the lab for many generations," Vera says. "The hope was that if we just kept giving it cereal leaf beetle eggs to attack, after many generations it might become so accustomed to them that it would attack no other kinds of eggs. I reared them for a long time but it didn't work out. When we switched to moth eggs or any other kind of insect eggs, *Trichogramma* went ahead and attacked them. So then we knew that if we wanted parasites to attack only the eggs of the cereal leaf beetle we had to depend on *Anaphes*."

But there was one important result of that experiment. Vera Montgomery, who had come to the lab in her middle age, knew that she could make a career for herself in biological control, and so did her superiors. She received time off from the lab to take science courses at the University of Indiana at South Bend. Later, the USDA paid her tuition while she took courses in entomology at Notre Dame.

"I made many trips into the fields to collect insects," she says. "I was able to relive the field trips I had made with my children when they were young. I collected two hundred and five different *families* of insects and sometimes I stayed up until two o'clock in the morning putting together my collections for school."

At the lab, Vera Montgomery's close observation of the insects' behavior enabled her to help write several important papers for scientific journals. She also took courses in photography so that she could learn to use cameras with a microscope and take sharp pictures of the tiny insects in the lab. She used some of these photographs to illustrate her papers.

Vera Montgomery and the other staff members at the Niles laboratory were meeting the challenge. Not all problems are solved, but the men and women there were about to create one of the most imaginative biological control programs in America's history.

10
PROBLEM SOLVING

Anaphes, the parasitic wasp that attacks the eggs of cereal leaf beetles, was easy to raise in the laboratory at Niles. It has a brief life span. Within nine days it develops from an egg into an adult that is less than one millimeter long. As an adult, it lives only two or three days in the lab—a short time, though long enough to mate and produce a new generation.

The problem, as we have seen, was not *Anaphes* but the cereal leaf beetle. The beetle simply took too long to rear. The entomologists at the lab wondered if they could get *Anaphes* to sting the eggs of another insect that would be easier to rear. Bug hunters in Europe had sometimes found *Anaphes* in a close relative of the cereal leaf beetle. Perhaps it would attack another host in the lab.

Nature has formed *Anaphes* to be usually host specific

—to prefer to lay its own eggs in those of the cereal leaf beetle. Since other parasites seldom attack the eggs of that beetle, *Anaphes* has that supply of food always available for its own young.

Entomologists collected the eggs of some American beetles that were related to the cereal leaf beetle. When an *Anaphes* adult was put in the case with a variety of "unnatural" hosts, it very often laid its eggs in one particular kind. The eggs it seemed to prefer were those of the three-lined potato beetle, which can be reared in only thirty to forty days.

The staff members at Niles decided to begin rearing large numbers of potato beetles. They were already growing many trayfuls of oats in the lab to feed cereal leaf beetles. Potato beetles, however, do not eat oats. They will eat jimson weed, a foul-smelling relative of tomatoes. Like most weeds, it is easy to grow, and so the staff members began to plant large quantities of jimson weed in trays.

Tom Burger and other scientists watched the experiment carefully. After a while they saw that *Anaphes* was not parasitizing as many of the three-lined potato beetles' eggs as they had hoped. They took pictures through a microscope of *Anaphes* on different kinds of eggs.

The pictures were greatly enlarged and they told an interesting story. When *Anaphes* stung a cereal leaf

Scientists in the lab grow large numbers of three-lined potato beetles on a relative of the tomato called jimson weed.

beetle's egg, its ovipositor remained firm and straight. But when it stung a potato beetle's egg, the ovipositor bent under the strain. Obviously, *Anaphes* had trouble getting into the potato beetle's eggs.

The staff members knew that this beetle's eggs are naturally covered with a sticky substance. They decided to try to wash the eggs before giving them to *Anaphes*. After several experiments Burger and his colleagues developed a mixture of mineral spirits and a detergent that cleaned the eggs. From that time on, *Anaphes* easily parasitized the potato beetles' eggs and the lab was able to rear many thousands of these parasitic wasps.

Even then, there was another problem to solve. When the U.S. government learned that many detergents were polluting rivers and streams, it ordered the manufacturers to make changes in their product. The new detergents broke down in sewage treatment plants and were no longer a serious pollutant. Unfortunately, the new detergent ruined the beetles' eggs, and the lab lost a whole generation of potato beetles before they could find a new solution in which to wash the eggs.

The *Anaphes* program has been very successful. The parasites raised at Niles, and at Purdue and Michigan State universities, were released in a number of places throughout the Midwest. This parasite not only survived harsh American winters in the wild, but quickly spread several hundred miles into new areas. *Anaphes* is now thriving even in Canada, where it has never been released. Wherever the cereal leaf beetle goes, *Anaphes* is usually close behind. Studies by Tom Burger and his assistants showed that these little wasps destroyed seven out of every ten beetle eggs in some fields.

"We still had to go on raising cereal leaf beetles in the lab," Tom Burger says. "We were afraid that after four or five generations, *Anaphes* would get hooked on the eggs of the potato beetle. So every once in a while we slip in some eggs of the cereal leaf beetle."

Besides *Anaphes*, the lab was responsible for rearing the three tiny wasps—*Diaparsis*, *Tetrastichus* and *Lemophagus*—that attack the larvae of cereal leaf beetles. But problems developed. Crowded together in the lab, many of those wasps died. Those that did grow into adults apparently did not like life in captivity and refused to attack the beetles' larvae.

The staff members decided to rear the three larval parasites outdoors. They set up insectaries in fields in several places throughout the Midwest. The larval parasites became numerous in those fields, which were densely planted with oats and other grain crops.

"It takes at least three years to get a field insectary started," Tom Burger says, "but the parasites do a terrific job once they start to build up their populations. They keep the cereal leaf beetle population so low that we have to bring in new supplies of beetles to keep the whole thing running."

There are now field insectaries in a number of Midwestern and Northeastern states. During one recent year, nearly a half million parasitic wasps were released in commercial grainfields. To prove the program's success, workers must also collect some of the eggs and larvae from different grainfields and take them back to the Niles lab. There, the staff members examine them to see how many of them have been parasitized and by which parasites.

At first, this was an expensive and time-consuming process. It was hard to identify the parasites inside the hosts' eggs or larvae because they were so tiny and they were all at different stages of growth.

Vera Montgomery, working with Penelope DeWitt, a young colleague at the Niles lab, made an important contribution. They examined the larval parasites in all stages of growth and wrote an illustrated paper which helps other entomologists to identify them quickly. The paper was published in a well-known scientific journal.

The larval parasites have become an effective enemy of the cereal leaf beetle in this country. The most successful is *Tetrastichus*. This wasp quickly followed the beetles into the Northeast and has also reached Canada. In some fields which are heavily infested by cereal leaf beetles, *Tetrastichus* has parasitized nine out of every ten larvae.

"In many places these parasitic wasps can keep cereal leaf beetles from doing a lot of damage," Tom Burger says. "Sometimes farmers will have to use pesticides if beetles get out of control. But they will have to spray carefully, because if the parasites get wiped out in an area, the farmer could be in worse trouble than before. These parasitic wasps are the farmer's friend."

The cereal leaf beetle program is important for a number of reasons. It proves that biological control is possible over a wide area of the country. It saves the farmer money. It saves the environment from being drenched with poisons. And it proves how much a few people can do—even working with very little money—once they believe they can find a better way to solve a problem.

11
OTHER BIOLOGICAL METHODS

After centuries of experiments and some impressive achievements, science is at last convinced that biological control can work. Humanity can show great inventiveness in finding ways to get rid of pests. It is now clear how predators and parasites play their part. Here are some other promising ways.

• *Insect Diseases.* This is a form of biological control because disease is another "natural enemy" of living things. Entomologists fight insect pests by distributing disease organisms among them. These diseases strike only the pests and cannot be caught by human beings or other animals.

For instance, an insect called the cabbage looper is often attacked by a virus. The looper dies and its body dissolves on cabbage leaves, leaving the virus behind.

Scientists tell us that this virus is not killed during the making of coleslaw from cabbage. The average bowl of coleslaw we eat contains about four billion live particles of the cabbage looper virus! But it is harmless to humans.

The case of the Douglas-fir tussock moth illustrates how entomologists use insect diseases. The caterpillars of this moth feed on the needles of evergreen trees in the Northwest. For long periods of time the tussock moths do little damage. Every once in a while they have a population explosion and damage evergreens that are valuable to the wood products industry.

Foresters tried to control these population explosions with DDT and other pesticides, but they simply became more frequent. A few foresters noticed, however, that in some areas the population explosion was halted not by pesticides but by viruses and other diseases.

Now scientists are able to cultivate strains of this virus in their laboratories. They are able to manufacture it in a mixture that can be sprayed on trees just like a pesticide. When the caterpillars eat the virus-coated needles, they pick up the virus, spread it to other caterpillars, and the population collapses.

• *Cultural Controls.* "Cultural" is used here in the sense of cultivation. In this method, farmers change their own habits and environment to discourage pest insects. For instance, farmers change the time when they plant their crops so that the crops grow while the pest is neither

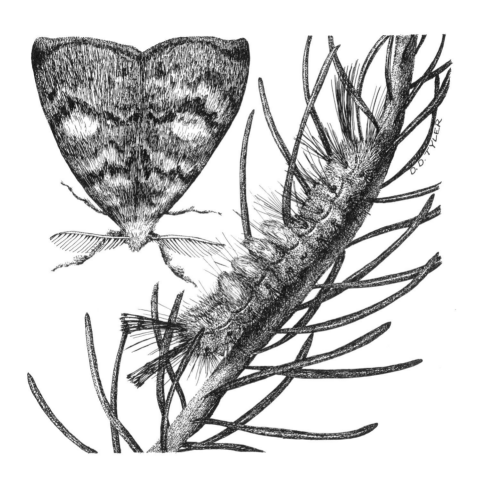

*The Douglas-fir tussock moth (above). When its larvae
(below) feeds on the needles of Western evergreens, it is
killed by a virus.*

active nor numerous. Farmers who plow the land deeply after harvest often uncover pests that like to spend the winter hiding in the soil. Other farmers either destroy their dead plants after harvest or plow them under the soil to get rid of pests which like to bore into old plant stems for the winter.

• *Resistant Plants.* Plant scientists are able to develop new strains of plants that resist attacks by pest insects. In Michigan they have developed a strain of wheat with hairy leaves that keep the cereal leaf beetle's larvae from feeding on them. Other scientists have developed wheat plants with solid stems. Certain insect pests like to lay their eggs only in hollow stems.

Scientists at Cornell University in New York discovered another defense that plants use against pests. There is a leafhopper that attacks bean plants. The scientists found that tiny hooked hairs on the leaves of certain bean plants caught many leafhoppers. The insects could not get free from the sharp hooks and died. If other bean plants could be developed carrying these hooked hairs, they would be well protected from leafhoppers.

• *Insect Pheromones.* The females of many kinds of insects have a natural perfume called a pheromone that attracts only the males of their own species. When the males are looking for mates they search for this odor, which leads them to the females.

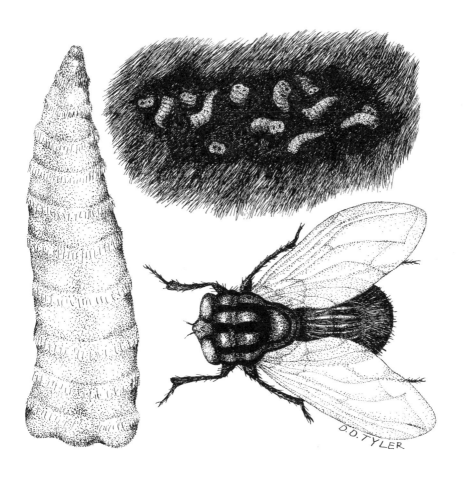

The screwworm is the larva of a fly that lays its eggs in the wounds or sores of large animals such as cattle.

Scientists now are able to make these pheromones in their laboratories and use them to get rid of pests. In controlling the gypsy moth, scientists have baited traps with pheromones and attracted males to their death. Other scientists have sprayed large amounts of the pheromones through fields and forests—this confused the males so that they were not able to find the females.

● *Sterilized Insects.* One of the great pests of cattle in Southern states was the screwworm. The adult of this species is a metallic-green fly that lays its eggs in the open wounds or sores of mammals. The eggs hatch into worm-like maggots that burrow into the flesh of cattle and other mammals, causing illness and even death.

Entomologists at the USDA developed a way to raise male screwworm flies and sterilize them by radiation. Then they released the flies. When the flies mated with females, which mate only once in their lives, the eggs did not hatch. Because of this program, the screwworm completely died out in Florida, and is under control now in large cattle-raising areas of the Southwest.

The USDA has also released millions of sterilized pink bollworm moths from planes over Florida's Everglades. Entomologists hope to kill off this pest that breeds in wild cotton plants in the Everglades and then spreads to cotton farms nearby.

12

PROGRESS: A PRODUCT OR A PROCESS?

How is humanity—today—meeting the two great challenges of controlling destructive pests and keeping poisons out of the environment?

Not very well despite advances in biological control. The chemical industry still tells us how to act. When a farmer has a pest problem, he often goes to a chemical salesman for advice rather than to an ecologist. That makes no more sense than if a sick person went to a drug salesman instead of to a doctor. The result is that the United States uses over one billion pounds of pesticides a year.

The salesman thinks that when he tampers with nature he does only *one* thing. The pesticide he sells is meant to do only one thing—kill pests. An ecologist, a scientist who studies the relationships among living things, knows that when you tamper with nature you never do only *one* thing.

For instance, when DDT was sprayed in a South American town some years ago to kill mosquitoes, the local cats were poisoned too. Small rodents that had been kept away from the town by the cats invaded all the houses. They brought with them fleas that spread a dangerous disease and three hundred people died. The men who used DDT thought they were only killing mosquitoes, but it is never that simple.

Biological control experts are truly ecologists. They know that even pests are part of the living community. Nature is complex, and ecologists understand that solutions to environmental problems must be complex too.

The chemical salesman tries to solve the problem of destructive pests by using a *product*—a pesticide. The biological control expert tries to solve the problem by using a *process*—putting nature to work through the use of natural enemies to control the pests.

Biological control is a fascinating science. It is as old as ancient China, and when given a chance it works. Rachel Carson, Paul DeBach, Vera Montgomery, and

---

The dwarf mistletoe is a plant that grows as a harmful parasite on the branches of other trees. Scientists infect the dwarf mistletoe with a fungus that often kills it.

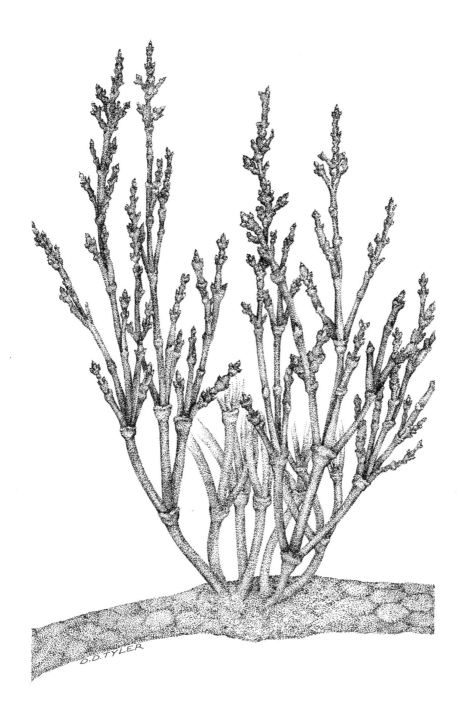

The ladybird beetle is a predator that eats many pest insects.

other scientists have made rewarding careers by trying to solve these complicated problems of life. The process of finding a solution pleases the mind, like solving a hard math question or a tricky chess problem.

Until now the answer to pest insects has been chemical pesticides, because people can make money selling pesticides. There is money to be made in a product, but not in a process. Not enough money has been put into biological control.

Scientists hope that this will change. In a world where millions of people are threatened with starvation, we have to find better ways to keep pests from destroying our crops.

Chemical pesticides will be used for some time to come because the natural enemies of pests have been destroyed in many places. Also, a few pests may not be as easy as others to control with natural enemies. But if we are to keep a healthy environment we must look for help from the scientists who use parasites, predators, and insect diseases in the process of biological control.

In the long run biological control is our best and safest weapon against pests. Otherwise, by flooding the environment with poisons, we may do more harm to ourselves and to our friends in the natural world than to the insects that some people think are our enemies.

As a wise scientist has said, "The object of our game with nature is not to win, but to keep on playing."

ACKNOWLEDGMENTS

The following books were extremely useful to us in our research:

Borror, Donald J., and White, Richard E. *A Field Guide to the Insects of America North of Mexico*. Boston: Houghton Mifflin, 1970.

Carson, Rachel. *Silent Spring*. Boston: Houghton Mifflin, 1962.

DeBach, Paul. *Biological Control by Natural Enemies*. New York: Cambridge University Press, 1974.

Evans, Howard E., and Eberhard, Mary Jane West. *The Wasps*. Ann Arbor: University of Michigan Press, 1970.

We would like to thank Thomas L. Burger, Vera Montgomery, and Murray Pender of the United States Department of Agriculture for sharing their experiences of working in biological control programs.

—ADA AND FRANK GRAHAM

INDEX

95

Index

game birds, 37

grain crops, 51, 53, 54. *See also* cereal leaf beetle

Great Lakes, 12, 52

grebes, 37–8

green lacewing, **66**

gypsy moth, 24–32; arrival of, 24–5; biological control of, 30–2, 44–5; DDT and, 34–5, 38; life cycle of, 25–8; natural enemies, 30–2; pheromones, 86; spread of, 28–9

honey bee, 12

houseflies, 45–6

hummingbirds, 10

Hymenoptera order, 60

India, 13

Indiana: cereal leaf beetle, 1, 3–4, 51

insectary: defined, 6; establishing field, 78–9

insecticides, *see* pesticides

insect pests, *see* pests

insects, 9–12; age of, 11; diseases, 81–2; insect-eating, 10–11; kinds of, 12; leaf-eating, 10; metamorphosis, 11–12; sterilization, 86; USDA war against, 34–42; word, 11. *See also* biological control; pests; predators and parasites

Italy, 56

Japan, 25, 30

jimson weed, 75

Kennedy, John F., 42

Klamath weed, 14

Koebele, Albert, 20–1

ladybird beetle, **90**

leaf-eating insects, 10

leafhopper, 84

lice, 33

locusts, 18; red, 13

Long Island, N.Y., 34

Massachusetts: gypsy moth in, 24–9

metamorphosis, 11–12

Michigan, 51–2, 83; Cereal Leaf Beetle Parasite Rearing Laboratory (Niles), 64–80

Michigan State University, 48–50, 64, 77

Midwest, 3–7; cereal leaf beetle, 3–7, 51–4, 62–4, 71, 77–8

mites, 11, 46; spider, 48–50

monarch butterfly, 12

97

wasps (*continued*)
68–78; life cycle, 60, 69; mud dauber, 60–1; paper, 60; parasitic, 61, 68–80; *Tetrastichus*, 68, 78, 79; *Trichogramma*, 71–2; wood, 13–14

wheat, 83
white-footed mouse, 30–1
woodcock, 37
wood wasp, 13–14
World War II, 33

Yugoslavia, 56